FREE DVVD

MW00908361

Essential Test Tips DVD from Trivium Test Prep

Dear Customer,

Thank you for purchasing from Cirrus Test Prep! Whether you're looking to join the military, get into college, or advance your career, we're honored to be a part of your journey.

To show our appreciation (and to help you relieve a little of that test-prep stress), we're offering a **FREE** *MTEL Biology Essential Test Tips DVD** by Cirrus Test Prep. Our DVD includes 35 test preparation strategies that will help keep you calm and collected before and during your big exam. All we ask is that you email us your feedback and describe your experience with our product. Amazing, awful, or just so-so: we want to hear what you have to say!

To receive your **FREE** *MTEL Biology Essential Test Tips DVD*, please email us at 5star@cirrustestprep.com. Include "Free 5 Star" in the subject line and the following information in your email:

1.   The title of the product you purchased.
2.   Your rating from 1 – 5 (with 5 being the best).
3.   Your feedback about the product, including how our materials helped you meet your goals and ways in which we can improve our products.
4.   Your full name and shipping address so we can send your **FREE** *MTEL Biology Essential Test Tips DVD*.

If you have any questions or concerns please feel free to contact us directly at 5star@cirrustestprep.com. Thank you, and good luck with your studies!

* Please note that the free DVD is <u>not included</u> with this book. To receive the free DVD, please follow the instructions above.

MTEL Biology (13) Rapid Review Flash Cards Book

TEST PREP INCLUDING 350+ FLASH CARDS FOR THE MASSACHUSETTS TESTS FOR EDUCATOR LICENSURE

Table of Contents

Introduction

Congratulations on choosing to take the Massachusetts Tests for Educator Licensure (MTEL) Biology (13) test! By purchasing this book, you've taken the first step toward becoming a biology teacher.

This guide will provide you with a detailed overview of the MTEL Biology test, so you know exactly what to expect on test day. We'll take you through all the concepts covered on the test and give you the opportunity to test your knowledge with practice questions. Even if it's been a while since you last took a major test, don't worry; we'll make sure you're more than ready!

WHAT IS THE MTEL BIOLOGY TEST?

The MTEL Biology test measures aptitude in biology for teacher candidates looking to certify as biology teachers. This test must be taken *in addition to* the assessments in reading, writing, mathematics, and professional knowledge required for all individuals seeking certification in Massachusetts. The MTEL Biology exam does not replace these other exams.

WHAT'S ON THE MTEL BIOLOGY TEST?

The MTEL Biology test gauges college-level content knowledge in biology and life sciences, as well as the necessary skills for biology. Candidates are expected to demonstrate thorough and extensive conceptual knowledge of subjects including cell biology, organismal biology, human anatomy and physiology, genetics, evolution, and ecology. You will also be expected to demonstrate mastery of key skills including scientific inquiry, research, and real-world connections. The content is divided into seven subareas, further divided into a total of twenty-three objectives.

You will have four hours to answer 100 multiple-choice questions and complete two open-response assignments.

What's on the MTEL Biology (13) Test?

Subarea	Objectives	Percent
I. Nature of Science	1. Scientific inquiry and scientific processes	10%
	2. Collection, organization, analysis, and reporting of data	
	3. Scientific instruments, materials, and safety practices	
	4. Historical and contemporary relationships among science, technology, and society	
II. Chemistry of Life and Cell Biology	5. Chemical components of living systems; basics of biochemistry	14%
	6. Cell structure and function	
	7. Physiological processes of cells	
III. Characteristics of Organisms	8. Structures, organization, and processes of plants	14%
	9. Structures, organization, and processes of archaea, bacteria, protists, fungi, and invertebrates	
	10. Structures, organization, and processes of vertebrates	
IV. Human Anatomy and Physiology	11. Structures and functions of the human digestive system; nutrition	14%
	12. Circulatory and immune systems	
	13. Respiratory and excretory systems	
	14. Nervous, endocrine, and reproductive systems	
	15. Skeletal, muscular, and integumentary systems	
V. Genetics, Evolution, and Biodiversity	16. Principles of heredity	14%
	17. Molecular basis of genetics	
	18. Theories and mechanisms of evolution	
	19. Biodiversity and classification	

Subarea	Objectives	Percent
VI. Populations, Ecosystems, and the Environment	20. Populations, communities, ecosystems, and biomes 21. Cycling of materials and flow of energy in ecosystem 22. Effects of human activity on the environment	14%
VII. Integration of Knowledge and Understanding	23. Prepare an organized, developed analysis on a topic from one of the preceding subareas	20%

Subarea I emphasizes scientific processes. Specifically, this includes formulating and investigating scientific questions, and the proper design and implementation of experiments, from proper handling of data to mastery of key equipment. This subarea also assesses your understanding of the larger context of science from both the historical and current perspectives.

Subarea II assesses your understanding of the chemical aspects of biology as well as your understanding of cellular biology. You must demonstrate mastery of the key molecules and chemical processes involved in life. You should also be able to identify the structures and functions of cells, compare different types of cells, and explain their major processes.

Subarea III assesses your understanding of all living things including plants, archaea, bacteria, protists, fungi, invertebrates, and vertebrates. You must be able to explain the life cycles, reproductive strategies, and significant processes of each category of life. You must also be able to identify their major classifications and how the living creatures maintain homeostasis.

Subarea IV focuses on human anatomy and physiology, including the location of bones, joints, and major organs. You should also know how joints, muscles and skin function, and the body parts and processes involved in the endocrine, reproductive, excretory, respiratory, nervous, and circulatory systems.

Subarea V assesses your understanding of genetic processes. You must demonstrate an understanding of the structure of nucleic acids and chromosomes and how genetic information is transferred. You also should understand differentiation and specialization of cells, and the general nature of mutations. Finally, you must demonstrate a mastery of evolution. Topics may include sources of genetic variation, mechanisms of evolution, supporting evidence, and significant scientific explanations for the origins of life, speciation, and extinction.

Subarea VI covers the relationships between living organisms and the environment and the role of those relationships in shaping specific ecosystems. You should know the relationships within and among species and the factors that influence population size. You should also understand the changes that occur during ecological succession, the different types of biomes and how energy flows within them, and the biogeochemical cycles. Finally, you should demonstrate an understanding of how natural phenomena can impact biodiversity and ecosystems, how ecosystems connect to each other, and how humans impact ecological systems.

Subarea VII assesses your ability to apply your knowledge and express it in a coherent and organized piece of writing. Each question may address a single aspect of one of the subareas or ask you to draw on multiple subareas.

How is the MTEL Biology Test Scored?

You will receive your scores on your MTEL Biology test on a predetermined release date associated with your testing window. Depending on when you test within the window, this could range from three to seven weeks after your test date. For more information, check http://www.mtel.nesinc.com. Your MTEL scores will automatically be sent to the Department of Elementary and Secondary Education and added to your licensure file once you apply for a license; however to have them sent to a college or university, you must make a request when you register for the exam. You can choose to either have the scores emailed to you or posted to your account.

Each multiple-choice question is worth one raw point. The total number of questions you answer correctly is added up to obtain your raw score. Constructed responses are scored holistically by two separate graders who are unaware of each other's evaluations of your work. If they differ substantially, to ensure fairness a third scorer determines your final essay score. This number is added to your raw score which is then converted to a scale of 100–300. Constructed responses are scored on a scale ranging from 1 to 4 on four criteria: purpose, subject matter knowledge, support, and rationale.

The multiple-choice section constitutes 80 percent of your score, and the constructed responses 20 percent. In order to pass the MTEL Biology test, you must receive a score of at least 240.

There will be some questions on the test that are not scored; however, you will not know which ones these are. MTEL uses these to test out new questions for future exams.

There is no guess penalty on the MTEL, so you should always guess if you do not know the answer to a question.

How is the MTEL Biology Test Administered?

The MTEL Biology test is a computer-based test offered in testing windows throughout the year at a range of universities and testing centers. Check https://www.mtel.nesinc.com/ for more information.

You will need to print your registration ticket from your online account and bring it, along with your identification, to the testing site on test day. No pens, pencils, erasers, printed or written materials, electronic devices, or calculators are allowed. You also may not bring any kind of bag or wear headwear (unless for religious purposes). You may take the test once every forty-five days.

About Cirrus Test Prep

Cirrus Test Prep study guides are designed by current and former educators and are tailored to meet your needs as an incoming educator. Our guides offer all of the resources necessary to help you pass teacher certification tests across the nation.

Cirrus clouds are graceful, wispy clouds characterized by their high altitude. Just like cirrus clouds, Cirrus Test Prep's goal is to help educators "aim high" when it comes to obtaining their teacher certification and entering the classroom.

About This Guide

This guide will help you master the most important test topics and also develop critical test-taking skills. We have built features into our books to prepare you for your tests and increase your score. Along with a detailed summary of the test's format, content, and scoring, we offer an in-depth overview of the content knowledge required to pass the test. Our sidebars provide interesting information, highlight key concepts, and review content so that you can solidify your understanding of the exam's concepts. Test your knowledge with sample questions and detailed answer explanations in the text that help you think through the problems on the exam and practice questions that reflect the content and format of the MTEL Biology test. We're pleased you've chosen Cirrus to be a part of your professional journey!

Molecular and Cellular Biology

adhesion

enzyme

Adhesion is the property of water that attracts it to other molecules. Adhesive forces will pull apart the individual atoms within a water molecule and pull them towards other molecules.

An enzyme is a protein produced by the cells of living organisms. It functions as a catalyst, accelerating or instigating specific biochemical reactions within an organism.

fermentation

inorganic compound

eukaryote

Fermentation is a biochemical reaction that breaks down sugars and releases energy without the use of oxygen. There are two types of fermentation: alcoholic fermentation and lactic acid fermentation.

Inorganic compounds are compounds that do not contain carbon, which is considered a building block of life.

Eukaryotic cells are cells that contain membrane-bound organelles, including a true nucleus enclosed by a nuclear envelope and a mitochondrion that acts as an energy-producing powerhouse of the cell.

cellulose

selective permeability

diffusion

Cellulose is a complex carbohydrate that makes up the majority of the structure in the cell walls of plants. Cellulose is the primary building material of plants, giving structure and integrity to plant stems, leaves, and roots.

Selectively permeable membranes regulate equilibrium within a cell by allowing some substances to pass through the membrane, while other substances are prevented from doing so.

Diffusion is the process by which molecules move from areas of high concentration—areas that contain large numbers of particles—to areas of low concentration.

pinocytosis

interphase

cytokinesis

Pinocytosis is a form of endocytosis in which cells engulf extra-cellular fluid that is not permeable through the cell membrane.

During the interphase stage of the cell cycle, the cell performs its regular functions to sustain life— such as cellular respiration—while also doubling in size and duplicating its DNA in preparation for division.

Cytokinesis is the division of a parent cell's cytoplasm that occurs after mitosis is complete.

daughter cell

anaphase

G1 checkpoint

During mitosis, a parent cell divides into two genetically identical daughter cells, which contain the same type and number of chromosomes as the parent cell.

Anaphase is the phase of mitosis during which individual chromatids of each chromosome in the cell split into distinct chromosomes and migrate to opposite ends of the cell in preparation for division.

During the G1 checkpoint of mitosis, the cell assesses its size, nutrient and energy availability, positive molecular cues, and DNA damage. Then, the cells use this information to either allow themselves to continue with mitosis or prevent themselves from entering the S phase.

homolog

nucleotide

deoxyribose

A homolog is one of the two members of a chromosome pair, with one member inherited from the mother and the other from the father. They differ from sister chromatids because they have different alleles for the same gene.

Nucleotides are molecules that serve as the building blocks of nucleic acids (DNA and RNA). They are comprised of a phosphate group, a five-carbon sugar, and a nitrogenous base.

Deoxyribose is the sugar found in DNA. It has one less oxygen atom than its counterpart in RNA— ribose—and is therefore more stable in structure.

histone

Okazaki fragments

activator

Histones are proteins that are found in chromatin and function as spools around which DNA strands can wrap themselves. They organize DNA strands into structures known as nucleosomes.

These are short fragments that form on the lagging template strand of DNA during the replication process. They have base pairs that are complementary to the bases found on the leading template strand.

An activator is an example of a transcription factor that "turns on" a target gene by binding to the gene's DNA to allow RNA polymerase to attach and make a new RNA molecule from the DNA's information.

gene expression

stem cell

substitution mutation

Gene expression is the process by which genetic information encoded in DNA is activated, transcribed, and translated into proteins, which instigates a target effect in a gene.

A stem cell is an undifferentiated cell found in a multicellular organism. Stem cells are capable of being differentiated into multiple different kinds of cells.

During a substitution mutation, one DNA base is inserted into the incorrect place instead of the intended base pair.

germline mutation

mutagen

polymerase chain reaction

A germline mutation is an inherited mutation that arises from alterations made to the sperm and egg cells; it is transmitted to offspring.

A mutagen is a physical or chemical agent that can alter genetic material and lead to mutation in an organism.

A polymerase chain reaction, or PCR, is used to make millions of copies of DNA segments by using primers to attach to strands of DNA.

gene therapy

recombinant DNA

carbon

During gene therapy, healthy genes are inserted into cells that have missing or abnormal genes in order to treat or correct genetic diseases and disorders.

Recombinant DNA is a DNA sequence that has been formed artificially by recombining genetic material from different organisms.

the atom that is found in all organic molecules and is necessary for life to function

glucose

rough endoplasmic reticulum

cytoplasm

the sugar molecule that serves as a primary energy source for most living things, is produced via photosynthesis, and is broken down during cellular respiration

the organelle within the cells of eukaryotes that produces proteins via ribosomes attached to its outer layer

the viscous fluid found within a cell that contains all organelles within a cell

phospholipid bilayer

flagellum

osmosis

the polar membrane that creates a barrier around cells by providing a hydrophobic interior layer and a hydrophilic exterior layer

a slender appendage that is made by the cytoskeleton, extends from the cell, and can whip back and forth to generate cell movement

the diffusion of water across a cell membrane

vesicle

lysosome

chromosome

a fluid-filled cellular structure found in cell membranes of eukaryotic cells that plays a major role in storing and transporting materials in and out of a cell

a membrane-bound organelle created by the Golgi apparatus and used to break down food material found in animal cells that contain enzymes

the strand-like structure of genetic material that carries the genetic information of the cell within the nucleus

centromere

telophase

S phase

the point of a chromosome where the two sister chromatids are linked together

the final phase of mitosis during which nuclear membranes form around the genetic material of the daughter cells

the cell cycle's specific stage of interphase during which the DNA of the chromosomes are duplicated

haploid

crossing over

DNA polymerase

cells that contain a single set of unpaired chromosomes, or half the amount of chromosomes of the parent cell

the stage of meiosis between prophase I and metaphase I during which homologous chromosomes exchange genetic information

the enzymes that build DNA molecules by assembling nucleo-tides during the process of DNA replication

leading strand

lagging strand

promoter

the strand of DNA found that is continuously being replicated during the elongation process

the strand of DNA that undergoes replication discontinuously in fragments during the elongation process

the region of DNA where RNA polymerase and other proteins bind in order to start the transcription process

messenger RNA (mRNA)

stem-cell niche

cell differentiation

the group of RNA molecules that delivers information transcribed from DNA to ribosomes for protein synthesis

the microenvironment within an area of specialized tissue that houses and maintains stem cells and regulates their fate

the process by which an unspecialized cell becomes specialized in order to perform a specific function

genetic variation

pluripotency

nondisjunction

the diversity of gene frequencies between individuals or between populations of organisms

the ability of a stem cell to give rise to multiple specialized cell types

the type of mutation that occurs when a chromosome fails to fully separate during meiosis, resulting in gametes with abnormal numbers of chromosomes

inversion

spectrophotometer

somatic cell nuclear transfer

a type of genetic mutation that occurs when chromosomal sections break off and reattach to the chromosome facing the opposite direction

an instrument used to measure very small sample sizes by shining beams of light through a prism in order to break it down into individual wavelengths, which are then shone through a liquid sample and quantified

the process of creating genetically identical organisms by transferring a nucleus to a different adult cell

What are the four major biological macromolecules?

Why is water necessary for cells to perform life functions?

What is the difference between an enzyme and a catalyst?

Carbohydrates, lipids, proteins, and nucleic acids are the four major biological macromolecules. Carbohydrates provide energy; lipids store energy and provide structure to cells; proteins perform chemical reactions; and nucleic acids provide genetic information for cells.

Water accounts for approximately two thirds of the material of every cell. All water molecules are comprised of two hydrogen atoms bonded to one oxygen atom. Water forms essential bonds with carbon, which is the element necessary for life and the site of all chemical reactions within a cell.

A catalyst is any substance that can instigate or accelerate a chemical reaction while remaining unchanged. An enzyme is a specific type of catalyst that accelerates chemical reactions among biological macromolecules. Enzymes are specialized to catalyze specific reactions within a living organism.

What is the function of chlorophyll in photosynthesis?

What are the major stages of cellular respiration?

What is the difference between chemosynthesis and photosynthesis?

Chlorophyll is a pigment found within a chloroplast—a specialized organelle found in plant cells. Chlorophyll absorbs and traps light energy, which is then used by the chloroplasts to provide energy to break down carbon dioxide and water in order to produce glucose.

Glucose and oxygen enter the cell in the initial stage of cellular respiration; glucose is then broken down into smaller molecules. The molecules and the oxygen next move into the mitochondria, where the molecules combine with the oxygen to further break down glucose and release hydrogen. Mitochondria release chemical energy to be delivered to other parts of the cell; they then release excess heat, water, and carbon dioxide as byproducts.

Chemosynthesis is the process of creating organic molecules—such as glucose—without the use of light energy; chemosynthesis uses chemical energy from inorganic compounds, such as sulfur. Photosynthesis relies on light energy from the sun.

Why are cell membranes composed of lipids?

Why are bacteria considered prokaryotic cells?

Which organelles differentiate plant cells from animal cells?

Lipids are generally composed of fatty acids and glycerol. Their chemical arrangement prevents them from mixing with water. This allows the lipids within the cell membrane to maintain and regulate the barrier between water within the cell and water outside the cell.

Bacteria cells, along with archaea cells, are single-celled organisms that do not contain a true nucleus or membrane-bound organelles. The term prokaryotic is derived from Greek terminology that means before nucleus.

Plant cells contain organelles that are not found in animal cells: chloroplasts, which are required for photosynthesis; cell walls, which provide plant cells with a rigid structure; and vacuoles, which are fluid-filled sacs in the cytoplasm that store nutrients and waste.

How does a cytoskeleton provide a cell with shape and structure?

What is the difference between active and passive transport?

What role do proteins play in cell communication?

A cell's cytoskeleton is comprised of a network of filaments and microtubules. These filaments and microtubules are made of different types of proteins. The interlocking nature of these structures provides the cell with its shape and stability and can also expand or contract to change the cell's shape to enable movement.

Transport is the movement of materials in and out of a cell across a cell membrane. Passive transport moves materials without using cell energy, while active transport moves materials with the use of energy.

Proteins act as receptors and senders of cell signals, passing and modifying the signals as they move throughout the cell to their target destination on the signaling pathway. Once the target protein receives the signals, it can trigger and direct the cellular response to the signals.

What role does diffusion play in photosynthesis?

What occurs in a cell if a particle is too large to be transported through the cell membrane?

Why do cells in multicellular organisms divide?

As photosynthesis produces oxygen molecules, the concentration of oxygen molecules within the plant cell becomes greater than the concentration of oxygen molecules outside the plant cell. This causes the oxygen molecules to move to the area of low concentration outside of the cell.

When a cell needs to absorb a particle that is too large to cross the cell membrane, it will undergo endocytosis—wrapping the membrane around the outside particle and moving the encased material within the cell. Similarly, a cell will perform exocytosis to rid itself of a large particle by wrapping the particle with a membrane; this membrane then attaches to the cell membrane and pushes the particle outside the cell.

Cell division occurs in multicellular, eukaryotic organisms in order to grow the organism's size, specialize in specific functions during the organism's development, and repair aged and damaged cells.

What is the function of a checkpoint in the life cycle of a eukaryotic cell?

What is the difference between mitosis and meiosis?

What occurs during the three main stages of interphase in the cell cycle?

During cell division, a eukaryotic cell regulates its division at multiple checkpoints in order to use internal and external environmental cues to decide whether or not to complete division. The three major checkpoints are the G1 checkpoint, the G2 checkpoint, and the spindle checkpoint.

Mitosis is the process of cell division in eukaryotic cells that results in two genetically identical daughter cells. Meiosis is the process of cell division in sexually reproducing organisms in which the cells divide twice to result in four sex cells that each contain half of the normal amount of chromosomes.

Three main stages of interphase are Gap 1 (G1), Synthesis (S), and Gap 2 (G2). Cell growth and protein production occur during G1; DNA duplication occurs during S; and cells continue their growth and protein production during G2 in preparation for mitosis.

What is the difference between meiosis I and meiosis II?

How are strands of DNA held together in a double helix?

What are the rules of base pairing in DNA?

During meiosis I, the parent cell contains two copies of each chromosome pair, or homolog. The cell divides into two daughter cells, which each contain two copies of one homolog of each pair. During meiosis II, the two daughter cells are divided again, this time with one set of chromosomes.

DNA molecules are comprised of two individual strands of nucleotides that are wrapped together like a ladder, known as the double helix. The nucleotides contain base pairs, which are held together to the base pairs of the other strand via hydrogen bonds.

DNA has four types of nitrogen bases: adenine (A), thymine (T), guanine (G), and cytosine (C). Adenine pairs with thymine, while guanine pairs with cytosine.

MOLECULAR AND CELLULAR BIOLOGY

What is the function of RNA?

What are the primary physical differences between DNA and RNA?

What are the stages of DNA replication?

Ribonucleic acid (RNA) molecules are responsible for synthesizing protein and transmitting genetic information. There are three major types of RNA: mRNA—messenger RNA that carries information from DNA; rRNA—ribosomal RNA that forms ribosomes for protein synthesis; and tRNA—transfer RNA that brings mRNA to the ribosomes.

DNA and RNA are nucleic acids that perform different functions within the cell. DNA is double stranded, while RNA is single stranded. DNA contains deoxyribose—a sugar—and has four bases: adenine, thymine, guanine, and cytosine. RNA contains ribose instead of deoxyribose, and has uracil as a base instead of thymine.

DNA replication occurs in three major stages: initiation, elongation, and termination. During initiation, the DNA double helix unwinds by breaking the hydrogen bonds between base pairs. During elongation, the new strand begins to grow and replicate according to base pairs. During termination, the two new double helices close off and repair any errors.

What are the stages of protein synthesis?

How do transcription factors help ensure that DNA is transcribed correctly during protein synthesis?

What is the purpose of RNA splicing?

Protein synthesis occurs in two major stages: transcription—during which messenger RNA copies genetic information from a DNA strand, and translation—during which mRNA information is decoded and used to guide the creation of new proteins.

Transcription factors are a wide variety of protein molecules that can either activate or repress the transcription of genetic material by binding the RNA polymerase to the DNA of the gene.

RNA splicing is a stage of mRNA processing. During this stage, noncoding regions of DNA—called introns—are removed from the chain and leave behind the regions that can be coded for protein synthesis, called exons.

What role do codons play in the translation process?

How does regulation of gene expression impact cell differentiation?

What role do adult stem cells play in cell differentiation?

Codons are sequences of nucleotides that correspond to a single amino acid. During translation, these codons are transcribed by messenger RNA (mRNA) and decoded by transfer RNA (tRNA) into a sequence of amino acids in the protein.

Gene expression regulation relies on internal and external factors and ensures that the correct genes are activated at the appropriate time during a cell's life cycle. When internal and external cues dictate the need for specialized cells, the appropriate genes in the cells are activated or repressed in order to trigger the cell to develop the specialized need.

Specialized tissues in multicellular organisms will contain many undifferentiated stem cells, which can be differentiated through gene expression to replace damaged or destroyed cells in the specialized tissue.

Why do stem cells undergo asymmetric cell division?

What is the most common source of gene mutation in DNA?

What is the purpose of gel electrophoresis?

Asymmetric cell division results in daughter cells that are not identical to one another. In stem cells, asymmetric division gives rise to a differentiated cell with the target genes expressed, as well as another undifferentiated stem cell, which can be divided again to specialize as needed.

The majority of gene mutation occurs when DNA is copied incorrectly during the replication process. A single change to one base in a nucleotide can have a dramatic impact on the organism and cause a disease or disorder.

Gel electrophoresis is performed to separate biological macromolecules—such as nucleic acids or proteins—based on their respective sizes.

What is the difference between reproductive and therapeutic cloning?

What occurs during the process of genetic engineering?

Reproductive cloning is the creation of a genetically identical organism from a host organism. In contrast, therapeutic cloning replicates specific genes, cells, and tissues for disease research and treatment.

Genetic engineering isolates DNA fragments from an organism by inserting the isolated fragment into a vector genome, with the growth, expression, and inheritance of the gene in the recipient.

2

Genetics and Evolution

trait

allele

In genetics, a trait is a characteristic or feature expressed in an organism. The type of trait expressed in an organism is called a phenotype and is determined by the DNA found in the genetic material for that particular trait.

An allele is one version of a pair of genes that is found on the same spot on a chromosome and controls the same trait in an organism. Individuals inherit one allele from each parent.

monohybrid cross

F1 generation

gene linkage

In a monohybrid cross, two individual parent organisms are selected for expressing the dominant and recessive versions of one particular trait and are intentionally bred to produce an offspring that carries genetic information from both parents.

The F1 generation is also called the first filial generation and is the first generation of offspring produced in a genetic cross experiment. Since the parents, or P generation, are selected for being homozygous for a particular trait, the offspring will be heterozygous for the selected trait.

Genes that are located in close proximity on a chromosome are considered linked because there is a high likelihood that the two genes will be inherited simultaneously in a resultant offspring.

cododominance

allosome

polygenic inheritance

Codominance occurs when there are more than two alleles for the same gene. If an offspring inherits two different dominant alleles from both parents, both alleles will be expressed in the offspring's phenotype.

An allosome refers to the *X* and *Y* chromosomes, which determine the sex of an offspring. All other nonsex chromosomes are called autosomes.

Polygenic inheritance occurs when multiple genes interact to determine the phenotype of a trait. Examples of polygenic traits include skin, hair, and eye color as well as height and weight.

modifier gene

uniparental inheritance

plasmid

Modifier genes alter the phenotypic expression of another gene by either diminishing or augmenting the expression of that gene.

Uniparental inheritance is a form of organelle inheritance that passes on genetic information from a single parent. During uniparental inheritance, genetic material found outside the nucleus is copied from the mother cell during fertilization. The resulting offspring does not obtain non-nuclear DNA from the father cell.

Plasmids are DNA-containing molecules that are separate from chromosomal DNA and replicate independently. They are typically small, circular, form in the cytoplasm of prokaryotes, and commonly used in genetic engineering techniques.

monoploid

genetic variation

migration

A monoploid organism contains half of the normal number of chromosomes in its adult state. Monoploids can live to adulthood, but are sterile and unable to reproduce.

Genetic variation is the diversity of genetic information found both within a population and between populations. It arises from mechanisms such as mutation, recombination, and genetic drift.

Migration is the movement of organisms from one population to another. Migration can lead to gene flow through the introduction of new alleles or the increase in frequency of alleles in the population, which can increase the diversity of the gene pool.

allele frequency

biological fitness

founder effect

Allele frequency is the rate of occurrence of a specific allele within the gene pool of a population. Allele frequency is determined by the use of the Hardy-Weinberg equation.

Biological fitness refers to the ability of an organism to successfully pass on its genes through reproduction compared to the success rate of other individuals in a population.

The founder effect is an example of genetic drift in which a small population is isolated from a larger population; as a result, the small population has decreased genetic diversity. The genes carried by the individuals in the small population get passed on at a proportionally higher rate than they do in the original population.

adaptive radiation

analogous structures

speciation

Adaptive radiation is a major mechanism of evolution in which a gene pool rapidly diversifies when a species is introduced to a new environment and must fill multiple new ecological niches.

Analogous structures are body features shared by two or more different species of organisms that are similar in terms of physical appearance and function but evolved from organisms that originated from different lineages. Analogous structures arise as a result of convergent evolution.

Speciation is the process by which a population of organisms diverges from its parent species and becomes genetically distinct.

habitat isolation

endosymbiosis

background extinction rate

Habitat isolation is a mechanism that keeps two populations apart from one another due to physical barriers between habitats or differing characteristics of habitats. Habitat isolation decreases the likelihood that populations of a species will interact and reproduce, which increases the chances that they will become genetically distinct and speciate.

Endosymbiosis is the theory that mitochondria and chloroplasts originated as free-living bacteria and evolved into energy-producing organelles in eukaryotic organisms millions of years ago. This theory is considered an example of evolutionary evidence.

The background extinction rate is a measurement of how often species become extinct during a period of time.

recessive

law of segregation

Punnett square

a descriptive term for an allele or gene whose phenotype is masked by a more dominant allele or gene in an organism; a trait that an organism does not express; the organism carries the phenotype's genetic information and passes it on to offspring

the Mendelian law that describes how allele pairs separate during meiosis, resulting in distinct haploid cells

the diagram that shows the probability that an offspring will inherit a specific genotypic combination from its parents

pedigree analysis

sex-linked traits

incomplete dominance

the study of gene inheritance through the use of a diagram of familial genetic history, which is used when a breeding experiment cannot be conducted

genetically linked traits that are located on either the *X* or *Y* chromosome of an individual organism

a form of inheritance that occurs when one allele is not completely dominant over another allele, which results in a phenotype that expresses elements of both alleles

epistasis

vegetative segregation

trisomy

this occurs when a modifier gene suppresses an expressed phenotype of another gene

the random replication of mitochondria, chloroplasts, and other organelles during cell divisions that results in daughter cells with random samples of these organelles

a form of nondisjunction that results in an extra chromosome being added to a cell during meiosis, leading to genetic disorders such as Down syndrome

translocation

hemophilia

gene pool

a chromosomal abnormality that occurs when nonhomologous chromosomes exchange information, which alters the structure of the chromosome

an *X*-linked recessive genetic disorder, expressed primarily in males, that occurs when a gene that produced blood-clotting proteins is deleted

the complete set of genetic information found within a population

GENETICS AND EVOLUTION 87

sexual selection

artificial selection

competitive exclusion principle

the form of mating during which organisms choose their mates based on advantageous or desirable traits

the mechanism of evolution that occurs when humans intentionally breed organisms with similar advantageous traits

the ecological concept that states that two species competing for the same limited resource cannot coexist and can lead to a species evolving distinguishing traits and behaviors

adaptation

gradualism

transitional fossils

an inherited trait that becomes common across populations or species due to natural selection and helps an organism become better fitted to its environment

the rate of evolution theory that states that evolution occurs as a result of many small changes taking place over long periods of time

fossilized remains that display characteristics of both ancestral groups and their descendants

post-zygotic isolation

allopatric speciation

law of superposition

a form of reproductive isolation that occurs after members of two different species have successfully reproduced, often resulting in decreased viability or fertility of hybrid offspring

a form of speciation that occurs as a result of geographic isolation between two species due to physical barriers

the law that states that younger strata of rock are found above older strata of rock, indicating that fossils found higher in rock strata are younger than fossils found further below

cephalization

mass extinction

What is the difference between
genotypes and phenotypes?

an example of an evolutionary trend in which nervous tissue concentrates in one end of an organism during embryonic development, eventually becoming a head

the loss of large numbers of species during a short period of time

A genotype is the combination of genes inherited from parents; a phenotype is the observable trait that is expressed in an organism.

Why do different combinations of alleles in an organism's genotype result in different phenotypes?

What are the laws of heredity that form the basis of Mendelian genetics?

What is the purpose of dihybrid crosses?

If an organism inherits identical alleles for a genotype from each parent, the genotype is considered homozygous, and the organism will express the same trait. If an organism inherits different types of alleles from each parent, then the genotype is considered heterozygous, in which case the dominant allele will be expressed while the recessive allele will not.

Mendelian genetics are founded upon three laws: the laws of segregation, independent assortment, and dominance. The law of segregation is based on the principle that genes separate into distinct alleles during meiosis. According to the law of independent assortment, these genes separate and recombine independently of one another, with every combination of alleles having equal chances to occur. The law of dominance states that offspring will express the dominant allele.

Dihybrid crosses are performed in order to isolate two different inherited traits from a set of parents to study how they are inherited and expressed through multiple generations of offspring. This helps determine whether or not any relationship exists between the two sets of alleles.

What do the squares and circles represent in a pedigree chart? What do filled and unfilled shapes represent?

How are homozygous and heterogeneous genes represented in a Punnett square?

What distinguishes Mendelian genetics from non-Mendelian genetics?

In a pedigree chart, males are represented by squares and females are represented by circles. Symbols that are fully filled in represent family members who are homozygous for a dominant trait; symbols that are partially filled in represent family members who are heterozygous for a dominant trait; symbols that are not filled in represent family members who are homozygous for a recessive trait.

Homozygous genes, which are genes that contain identical alleles from each parent, are referred to by two capital letters if the trait is dominant and by two lowercase letters if the trait is recessive. To contrast, heterozygous genes are referred to by a capital letter—representing the dominant allele—and a lowercase letter, which represents the recessive allele.

Mendel's laws of heredity apply to traits that are controlled by a single gene that has two potential alleles: one dominant and one recessive. Not all traits are controlled by these same patterns. Non-Mendelian genetics refers to patterns that are studied and observed in traits that do not have dominant/recessive allele relationships, traits that have more than two possible alleles, and traits that are controlled by multiple genes.

What occurs during the process of recombination?

Why are recessive X-linked genetic traits expressed primarily in males but not females?

Why are height and weight considered examples of traits that display continuous variation?

Recombination (crossing over) occurs during prophase I of meiosis. As duplicated chromosomes separate into chromatids, the chromatids intertwine and exchange alleles in random patterns. The resultant chromosome therefore carries the same genes but has a combination of both paternal and maternal alleles.

Females carry two *X* chromosomes, while males carry an *X* chromosome and a *Y* chromosome. If a female inherits a recessive *X*-linked allele, the phenotype has a greater chance of being masked by a dominant *X*-linked allele. Males inherit only one *X*-linked allele, so the allele will be expressed regardless of whether it is the dominant or recessive allele.

Unlike eye color—a trait for which there is not a set number of varieties—height and weight are traits that exist along a continuum.

How is extranuclear DNA passed on from parent to offspring?

How do prokaryotic organisms increase genetic variation during reproduction?

How do chromosomal abnormalities increase genetic diversity and drive evolution among a population of organisms?

Some organelles, including chloroplasts and mitochondria, contain DNA that does not replicate during gamete formation because it is not located within the nucleus. DNA in these organelles is duplicated from the maternal cell when two gametes fertilize, and the offspring inherits extranuclear DNA exclusively from the mother.

Bacteria and archaea increase their variation by exchanging genetic information through transduction, transformation, and conjugation. Transduction occurs when a cell is infected by a virus that absorbs genetic material and transfers it to another host cell. Transformation occurs when prokaryotic cells absorb genetic material from a source outside the cell. Conjugation occurs when two prokaryotic cells exchange plasmids, which are gene-carrying structures within the cell.

Some chromosomal abnormalities introduce new genes and changes to phenotypes that are expressed among a population of organisms. In sexually reproducing eukaryotes, these phenotypes can give organisms an ecological advantage and are more likely to be selected for and passed on in a population.

Why is cystic fibrosis considered an autosomal recessive disorder?

Which mechanisms allow alleles to be distributed and move within and between populations of species?

What are the five conditions of the Hardy-Weinberg equilibrium?

Cystic fibrosis is an inherited genetic disorder. It is the result of a mutation of an autosomal gene, meaning it is not linked to a sex chromosome. The mutated allele is recessive—the genes for cystic fibrosis will be masked in the presence of a nonmutated dominant allele. In order to inherit the disease, a human must receive a recessive allele from both parents.

Mechanisms for allele distribution and movement include gene flow, migration, genetic drift, and nonrandom mating.

The Hardy-Weinberg equilibrium, which states that allele and genotypic frequencies will remain in constant balance without outside evolutionary forces, is based on five conditions about the population: it is large, it is isolated from other populations, there are no genetic mutations, all mating is random, and there is no natural selection.

How do evolutionary mechanisms lead to change in populations of sexually reproducing organisms?

Which conditions must be met for natural selection to occur?

How do random events drive genetic drift in a population?

Evolutionary mechanisms, such as natural selection, alter the variation of genetic information and allele frequency within the population. These changes alter the reproductive patterns of a population; certain traits survive while others do not. This results in a gradual change to the population of organisms over time.

Within a population, natural selection requires that there is variation in traits that can be inherited by offspring; there is an overproduction of offspring (i.e., more are born than can survive); the offspring best suited for the environment are more likely to pass on their genes; and there is differential reproduction (i.e., not all members of a population will reproduce at the same rate).

Random events, such as natural disasters or disease, can wipe out large portions of an organism's population or isolate small populations from the larger population. This leads to a loss of diversity in the gene pool of these populations. This alteration to the relative frequency of alleles in a population is referred to as genetic drift.

How do ecological relationships lead to coevolution of two or more species?

What is the difference between convergent and divergent evolution?

What is the role of stasis in the theory of punctuated equilibrium?

During coevolution, two species with a close ecological relationship impact the evolution of the other species, and vice versa. This is seen among predator-prey relationships as one species evolves to catch or evade the other; among competitive species as they differentiate to occupy different niches; and among mutualistic species, which evolve mutually beneficial mechanisms in response to one another.

Convergent evolution occurs when two unrelated species evolve similar traits, or analogous structures, in response to similar types of environments. Divergent evolution occurs when two species descended from a common ancestor evolve nonsimilar traits in response to different environments or niches.

The theory of punctuated equilibrium is a model of evolutionary rate. The theory states that species remain in stasis—long periods of equilibrium with no major evolutionary changes—the majority of the time. This stasis is punctuated by short, rapid bursts of evolution that drive species change and speciation.

How does reproductive isolation play a
central role in speciation?

What are the different types of
pre-zygotic isolation?

Why do hybrid organisms experience
infertility?

Populations of species that become isolated from one another will see a reduction in gene flow between the two groups of organisms. Over time, different populations may undergo evolutionary changes that make the two populations incapable of inbreeding, which can lead to speciation.

Examples of pre-zygotic isolation include habitat, temporal, behavioral, mechanical, and gamete isolation. These types of isolation occur prior to copulation or fertilization, distinguishing them from post-zygotic isolation.

Hybrids are offspring that result from the successful reproduction of organisms from two distinct, yet similar, species. Hybrids that survive to adulthood often experience infertility because they do not have a full set of compatible chromosomes.

In the absence of physical barriers, which factors can lead to parapatric speciation?

Why do homologous structures in living organisms indicate evidence of evolutionary relationships?

What distinguishes evolutionary trends from random trait fluctuation?

Parapatric speciation is a rare form of speciation that occurs among populations that occupy the same habitats but become isolated due to changes in behavior or timing of breeding.

Homology of structural features between two species indicates that the species shared a common ancestor at some point in its evolutionary history. This is an example of one piece of evidence of evolutionary relationships between two species.

Evolutionary trends are changes to a trait or a structure that occur in one lineage or across the lineage of multiple organisms. These changes appear time and time again, with increasing quantities, sizes or complexities. In contrast, random fluctuation of traits does not show a distinct pattern across lineages, nor does it display increasing complexity over time.

3

Animals, Plants, and Ecology

taxonomy

heterotroph

Taxonomy is the scientific branch that classifies organisms into eight taxons based on their biological similarities and relationships.

Heterotrophy is a mode of nutrition that distinguishes different groups of organisms which obtain nutrients from external sources.

Monera

domain system

Protista

Monera was a kingdom found in the original five-kingdom classification system and consisted of prokaryotic, unicellular organisms. The kingdom has since been subdivided into two distinct kingdoms: Bacteria and Archaea.

the system of classification based on the molecular biology of organisms; this system replaced the traditional five-kingdom system as the preferred method of classification in the scientific community

the kingdom of classification that includes both unicellular and multicellular organisms with true membrane-bound nuclei

cell

acoelomate

endotherm

the simplest level of biological organization within an organism

Acoelomates are animals that do not have an internal body cavity that is lined with mesoderm. This group of organisms includes Porifera, Cnidaria, and Platyhelminthes.

Endotherms are animals that regulate their own body temperature by generating heat internally. This group primarily includes birds and mammals.

Porifera

segmentation

mantle

The Porifera phylum contains the sponges, which are characterized by their asymmetrical body plan, lack of a body cavity, and inability to move. Sponges are some of the simplest animal organisms and do not have specialized tissues or organs, although they do have specialized cells that perform different functions.

Segmentation is a feature found among several phyla of animals that also display bilateral symmetry. Segments are a series of repeated parts or compartments which repeat sets of structures along the body.

The mantle is a distinguishing feature found among the Mollusk phylum. It is a flat layer of tissue that covers and protects the soft-bodied Mollusk.

Actinopterygii

---

incomplete metamorphosis

---

radial symmetry

The Actinopterygii are a subclass of fish distinguished by bony skeletons and fins supported by ray-like structures; they are also the largest group of fish in the world.

Metamorphosis is the process by which an animal transforms from immature to adult form. During incomplete metamorphosis, insects undergo several gradual nymphal stages and do not have a pupal stage during which a complete transformation takes place.

the body plan found on animals whose body parts are arranged in a circle around a central point

parthenogenesis

sessile

Nematoda

the development of an embryo from an unfertilized female egg

an organism that is immobile, attached by its base to one spot

the phylum of roundworms, which are characterized by their bilateral symmetry, a pseudocoel, and a complete digestive system

exoskeleton

water vascular system

ovoviviparity

the external layer of chitin found in arthropods that offers protection and support to the body

the system of vessels found in the bodies of echinoderms that collects water from outside the body via tube feed and transports water throughout the body

the mode of reproduction in which eggs are developed and hatched inside the body of the mother

transpiration

stamen

endosperm

Transpiration occurs when water is released through open stomata and evaporates into the atmosphere.

The stamen is the male reproductive structure found in the flowers of angiosperms and is comprised of two main parts: the filament and the anther. The stamen is responsible for producing sperm cells and pollen grains.

The endosperm is a seed tissue that provides nourishment to the developing plant embryo contained in a seed.

zone of maturation

phellem

parenchyma

The zone of maturation is a section of a plant's root system where cells differentiate into specialized cells, such as vascular cells; it is located at the tip of the root.

Phellem, or cork tissue, is a layer of cells located in the periderm, or bark of the plant. Phellem is comprised of nonliving cell tissue and waxy suberin within the cell walls.

Parenchyma is a type of ground cell found in plants. It is an abundant, undifferentiated cell that can divide rapidly and specialize to fill multiple roles in the plant.

rhizoid

stomata

xylem

Rhizoids are root-like appendages found in the bryophytes, a nonvascular group of plants that includes mosses. Rhizoids absorb water from the soil and transport it from cell to cell.

the small pores found in the plant epidermis that allow water and gases to pass in and out of the plant

the plant tissue that is responsible for transporting water upwards from the roots to the rest of the plants using a series of tracheid cells and vessel elements

nonvascular plants

cones

cotyledon

the group of plants that lack xylem and phloem tissue, which is used to transport materials such as food and water throughout the plant

the tough scaly structures that enclose and protect the reproductive structures of gymnosperms

the seed leaf found in the seeds of developing plant embryos that store nutrients for future absorption; monocots have one of these, while dicots have two

metagenesis

allochory

biosphere

the life cycle of plants in which plants alternate between a diploid sporophyte generation and a haploid gametophyte generation; also known as alternation of generations

mechanisms of seed dispersal that depend on secondary agents, such as wind, water, and animals

The biosphere is the highest level of the ecological hierarchy and encompasses all living things and nonliving things on Earth. This includes land, air, water, and any other system in which life is found.

estuary biome

interspecific competition

carrying capacity

A biome is a vast series of ecosystems that are functionally similar to one another and feature comparable dominant species of plants and animals. Estuary biomes are found where the oceans meet fresh water and are characterized by constant water movement, which stirs up nutrient-rich material from the benthic areas and supports large amounts of vegetation and animal life.

Interspecific competition refers to two or more different species that occupy the same habitat. When their niches within the habitat overlap, there is increased competition for the same food, water, and space resources.

The carrying capacity is the point at which a population has reached the maximum size it can maintain in a given geographic area, assuming resource availability remains stable.

opportunist species

territoriality

decomposer

Opportunist species are species that tend to be small in size, reproduce rapidly, generalize rather than specialize, and have short life cycles. These characteristics enable these species to quickly take advantage of new environmental conditions.

Territoriality is a behavior displayed among some populations in a community and is used by an individual to defend a geographic area from both rival species and rivals within its own species. This includes calls, scent marking, intimidation displays, or attacking other individuals.

A decomposer is an organism that consumes dead or dying organisms, breaking down organic material into distinct inorganic components in the process. Examples of decomposers include bacteria, fungi, worms, and insects.

community

biomass

pyramid of energy

the collective population of all species that coexist in the same geographic area

the total mass of organisms and organic materials in a given geographic area

the chain of energy flow throughout an ecosystem in which there is a net loss of energy as it moves from one level to the next

biogeochemical cycle

nitrogen fixation

invasive species

the continual flow of chemicals between abiotic and biotic components within an ecosystem

the process in which nitrogen found in the atmosphere is converted to usable forms and deposited in the soil, where it becomes available to living things

non-native species that are introduced to new areas, either intentionally or unintentionally, and have a negative impact on ecosystem health

bioaccumulation

What characteristics distinguish living things from nonliving things?

What are the eight taxons of the domain classification system?

the process in which an individual consumes a pollutant faster than it loses the pollutant through metabolization or excretion, leading to increased amounts of the pollutant in the body

All organisms are made of organized cells; they all grow and reproduce; they all respond to the environment; and they all obtain and use energy to maintain life processes.

The eight taxons of the domain classification system are domain, kingdom, phylum, class, order, family, genus, and species.

How do unicellular colonies differ from multicellular organisms?

How do the different kingdoms differ in terms of reproduction and replication?

What is the order of biological organization in multicellular eukaryotes, from simplest to most complex?

A colony is a collection of multiple unicellular organisms living together in a mutualistic relationship; however, each cell performs all of its own life functions and can survive independently of the colony. In contrast, multicellular organisms have many specialized cells that are dependent upon one another for survival.

The Bacteria and Archaea kingdoms reproduce asexually through binary fission. The remaining kingdoms of eukaryotic organisms can reproduce either asexually—using binary fission, budding, fragmentation, sporulation, and other methods—or sexually, through the formation of gametes.

From simplest to most complex, the biological organization of multicellular eukaryotes ranges from organelles to cells, tissues, organs, and organ systems.

Why do more complex organ systems arise in Coelomates than in Pseudocoelomates or Acoelomates?

How do hermaphroditic organisms reproduce?

What are some advantages of ectothermy?

Coelomates have a coelom, a fluid-filled body cavity that is derived from the mesoderm, a germ layer that appears during embryonic development. Coeloms allow for separation and specialization of distinct organ systems through compartmentalization.

Hermaphroditic organisms contain both male and female reproductive systems. These organisms undergo sexual reproduction by either self-fertilization or by mating with another member of their species.

Ectotherms are organisms whose internal body temperature is determined by the external temperature of their environment. The majority of animal life is ectothermic. Since ectotherms do not use high amounts of energy to generate their own heat, they do not require as many food resources as endotherms. They can survive for long periods of time without eating and spend less energy searching for food; instead, the majority of food they consume provides energy for growth.

What are the two stages of the Cnidarian life cycle?

How do organisms that lack organ systems obtain oxygen?

How does the Annelida phylum differ from the other worm-like phyla, such as Nematoda and Platyhelminthes?

Cnidarians, which include jellyfish and corals, undergo two major stages: the sessile polyp stage and the motile medusa stage. During the polyp stage, free-swimming larvae become sessile and produce buds as a form of asexual reproduction. These buds are released as medusas, which are free swimming or floating.

Many phyla of animals do not have a true respiratory or circulatory system and instead acquire oxygen through air or water diffusion.

Annelids are the most complex of the three-worm phyla. Unlike Nematodes and Platyhelminthes, they display segmentation, a coelom, and a closed circulatory system.

Why is the circulatory system in Molluscs considered an open circulatory system?

What are the functions of the three main body segments in Arthropods?

What distinguishes a notochord from a backbone among members of the Chordata phylum?

In Molluscs, as well as Arthropods, blood pumps from a two-chambered heart directly into body spaces, where an exchange of materials occurs as blood surrounds tissues. Blood is not delivered to organs via vessels, as it is in a closed circulatory system.

Many Arthropods contain three primary body segments: the head—which contains the eyes, mouthparts, and antennae; the thorax—the middle section that contains the legs and wings, if wings are present; and the abdomen—which contains most major organs.

A notochord is a flexible, rod-shaped structure that extends from head to tail in Chordates. All chordates have a notochord during their embryonic stage of development, which may become reabsorbed or develop into a spinal cord. Not all Chordates develop a backbone, which is a series of vertebrae that protects the spinal cord.

Which groups of animals in the Chordata phylum lay amniotic eggs?

What are the parts and functions of plant leaves?

Why do many types of flowers contain layers of colorful petals?

Eggs that are amniotic are enclosed by fluid-filled membranes and a protective outer shell. Reptiles, birds, and some mammals lay amniotic eggs.

Leaves consist of blades and petioles. Blades are the broad sections of a leaf, with the upper side containing chloroplasts and initiating photosynthesis. Petioles are the thin part of the leaf that connect the blade to the stem and play a vital role in transporting sugars produced by photosynthesis to the stem for further distribution.

Large, colorful petals serve to attract potential pollinators, which are animals that play a vital role in plant reproduction by dispersing pollen grains to other flowers for fertilization.

What is the path water travels from the soil through the root system?

How does the phloem transport sugars from the leaves to the rest of the plant?

What is the difference between primary and secondary growth in a plant?

Water is initially absorbed by root hairs extending into the soil. It then passes through a layer of cells called the cortex, followed by the endodermis, and then the pericycle. After passing through these layers, water enters vascular tissue and is transported to other parts of the plant via the vascular cylinder.

Sugars are transported in phloem tissue through a series of sieve elements. These long, tubular cells are filled with sugars and water, then moved down the tube as the combined pressure of these substances forces the sugars out to other parts of the plant.

Primary plant growth refers to the growth that occurs in the roots and shoots of the plants, increasing the plant's height and extending its roots further into the ground. Secondary plant growth refers to the widening of the plant that occurs as vascular cambium produces more tissue.

What distinguishes ferns and fern allies from other types of vascular plants?

Why do dicots experience secondary growth, but monocots do not?

What are the stages of the alternation of generations?

Unlike gymnosperms and angiosperms, ferns and fern allies do not reproduce via seeds. Instead, they reproduce via spore cells.

Dicots are a group of angiosperms that contain vascular cambium, which is a tissue that gives rise to new plant tissue in the stems and leads to widening, or secondary growth, of the plant. Monocots do not contain vascular cambium.

Plants begin in a diploid form, or generation, called *sporophytes*. Sporophytes contain genetic information from both parents and undergo meiosis to produce haploid cells. These haploid cells divide during mitosis and become a new generation of plant, called a gametophyte. Through mitosis, these haploid gametophytes produce male and female gametes. After gametes are released and fertilization occurs, the resulting embryo becomes a sporophyte.

Which mechanisms do plants use to disperse seeds?

What is the function of plant meristem?

What are the differences between biotic and abiotic materials in an ecosystem?

Seeds are dispersed through several mechanisms: gravity, wind, water, internal animal, external animal, and self-ejection.

Meristematic tissues in plants contain undifferentiated cells that can give rise to multiple different types of cells in order to aid plant growth and replace destroyed or damaged cells.

Biotic materials are living things, or materials composed of once-living tissue, while abiotic materials are nonliving components of an ecosystem.

What are the major vertical zones of the ocean biome?

What are two defining characteristics of populations that are examined by the field of population ecology?

What is the difference between a habitat and a niche?

From the bottom to the top, the ocean biome is divided into three major vertical zones: the benthic zone, the pelagic zone, and the photic zone. The benthic zone is the ocean floor; the pelagic zone is the open ocean, and the photic zone is the surface layer closest to the sun.

Populations are studied by examining their density and dispersion. Population density is a measure of the number of individuals in a given unit of space, while dispersion is the pattern of how individual population members are spaced throughout a given area.

A habitat is the physical space in which an organism lives, while a niche is that organism's role and behavior within that habitat.

Compared to the exponential growth model, why is the logistic growth model appropriate for real-world populations?

What are the ecological advantages of asexual reproduction?

What are the three types of symbiotic relationships that occur between species?

Exponential growth models do not consider limiting factors, which are ecosystem components that can restrict or reduce population growth. Logistic growth models display a tapering of growth at the top of their curve, representing the impact of limiting factors on a population.

Asexual reproduction occurs at a rapid pace, which benefits a species when occupying new habitats, adapting to ecosystem change, or recovering from a population loss.

Symbiosis is an interdependent relationship between two species. Symbiotic relationships can be parasitic, in which one species benefits and the other is harmed; commensalistic, in which one species benefits and the other is unaffected; or mutualistic, in which both species benefit from the relationship.

What are the stages of primary succession
in an ecosystem?

Which types of organisms occupy the first
trophic level of an ecosystem?

Which level of a pyramid of biomass has
the least amount of biomass?

Primary succession begins with the pioneer plant stage, when simple, hardy organisms take advantage of newly available surfaces. Intermediate species follow the pioneer plants, which are bigger, more complex, and have longer life spans. This increases biomass in the area, and intermediate plants are continually replaced by more complex and mature plants until the ecosystem reaches its climate community stage, at which point it is dominated by mature plant life and experiences great species diversity and productivity.

The first trophic level of an ecosystem is filled by producers—organisms that produce their own food. This includes many groups of autotrophs, such as plants, plant-like protists, and bacteria.

Tertiary consumers, which occupy the highest trophic level of a biomass pyramid, have the smallest amount of biomass due to the fact that energy is lost as it moves up the trophic levels, and there is not enough ecosystem energy left to support higher numbers.

What is the difference between open and closed biogeochemical cycles?

How does the greenhouse effect lead to warming of the earth?

Some biogeochemical cycles, such as the hydrologic cycle, are considered closed because materials used by the organisms are recycled, not gained or lost. Open cycles, such as ecosystem energy cycles, can gain and lose elements.

Certain atmospheric gases, such as carbon dioxide and methane, are considered greenhouse gases because they trap thermal energy that is emitted from the earth's surface as well as heat energy emitted by the sun. This energy is radiated to the earth's surface and increases its temperature.

Human Anatomy and Physiology

blood pressure

capillaries

Blood pressure is the force exerted on the vessels of the cardio-vascular system as blood is pumped through the body.

Capillaries are the smallest blood vessels in the cardiovascular system. The thin network of vessels serves as the site of material exchange between arteries delivering blood and veins returning blood to the heart.

pulmonary ventilation

bronchiole

larynx

Pulmonary ventilation, or breathing, is the process of air filling the lungs during inspiration (inhalation) and releasing air from the lungs to the atmosphere during expiration (exhalation).

Bronchioles are small divisions of the bronchus, or air passage, found within the lungs. They connect to alveoli and transport gases in and out of the lungs during respiration.

The larynx is a hollow organ located in the neck on top of the trachea. In humans, the larynx contains vocal cords. As air passes over the larynx during respiration, the cords vibrate and create sound that can be manipulated to make speech.

chyme

bile

colon

Chyme is a mixture of partially digested food and digestive juices that forms in the stomach and is passed through to the small intestine for further digestion.

Bile is an acidic fluid that is produced by the liver and stored in the gall bladder. Bile aids in the breakdown of fat macronutrients during digestion.

The colon is a part of the large intestine that connects the cecum to the rectum. The colon is the largest part of the large intestine and is primarily responsible for water absorption and waste preparation.

nephron

spermatogenesis

menstrual cycle

Nephrons are small, fine tubes in the kidney that create urine and filter waste and other substances out of the blood.

Spermatogenesis is the process of sperm development in males. Spermatogonia, which are sperm stem cells, give rise to spermatocytes in the seminiferous tubules within the testes.

The menstrual cycle occurs in human females as well as other female primates. It is the process of ovulation (the release of the egg into the fallopian tube) and menstruation, which is the shedding of the egg and uterine lining in the event fertilization does not occur.

somatic nervous system

medulla oblongata

visceral muscles

The somatic nervous system is a part of the peripheral nervous system and serves to control voluntary movement of the skeletal and muscle systems.

The medulla oblongata is a portion of the hindbrain that connects the brain to the spinal cord and is primarily responsible for regulation of autonomic bodily functions.

Visceral (smooth) muscles are tissues found inside veins and arteries, the stomach, and other internal organs. These muscles involuntarily contract to move blood, food, or other substances from one place to another. They are so named because of the lack of striations, or bands, found in other muscles.

sarcomere

sesamoid bone

appendicular skeleton

The sarcomere is comprised of both thick and thin filaments and is the basic unit found in myofibrils, which are the elongated organelles found within striated muscle tissue. As the muscles contract, the thick filaments slide by the thin filaments, thus shortening the myofibrils and causing muscle contraction.

A sesamoid bone is a small, rounded bone or nodule that is found within a tendon where the tendon passes over a joint. The patella, or kneecap, is the largest example of a sesamoid bone in the human body.

The appendicular skeleton is the division of the skeletal system that contains the bones of the appendages, or limbs, as well as the pectoral and pelvic girdles that attach to the axial skeleton.

trophic hormones

melatonin

inflammation

Trophic hormones are hormones that are produced in order to stimulate other glands to produce hormones. Most trophic hormones are produced by the anterior pituitary gland.

Melatonin is a hormone produced by the pineal glands in order to control sleep and wake cycles.

Inflammation is a nonspecific response to pathogens and injured cells. Localized tissue releases histamines that raise the temperature and blood flow to the target area, which triggers more white blood cells to enter the area in order to repair it.

antibodies

hypodermis

melanocytes

Antibodies are proteins produced by plasma that bind to the antigen protein found on the surface of pathogens. Once attached, the pathogen becomes neutralized and phagocytes are stimulated to ingest the target cell.

The hypodermis is a deep layer of skin that stores fat, which provides insulation and support for the body.

Melanocytes are skin cells that produce melanin in order to protect the skin from UV radiation through sun exposure. The melanin pigment also gives skin its color.

thoracic duct

tonsils

atria

The thoracic duct collects lymph from the majority of the human body, then drains lymph into the bloodstream.

Tonsils are lymphatic nodules located in or near the pharynx. The human body has five tonsils which store *T*-cells and *B*-cells and aid the body in response to infection.

the chamber of the heart that receives blood from the lungs or other parts of the body

systole

external respiration

pleura

the contraction phase of the heartbeat, during which the heart ventricles force out blood into the lungs or to the body

the exchange of gases that occurs in the lungs at the alveoli

the delicate membrane that covers and lines the human lungs

peristalsis

absorption

esophagus

the muscular contractions that push food from one point to another within the digestive tract

the process of digested food moving into the bloodstream for cellular respiration

the part of the digestive tract that transports food from the throat to the stomach by contracting its muscular walls

sphincter

ureter

vas deferens

a circular muscle that, when relaxed, allows materials to pass through for excretion

a thin, muscular tube that transports urine from the kidney to the bladder

the long tube in the male reproductive system that serves as a pathway for sperm to travel to the urethra and mix with nutrients and fluids to form semen

ectoderm

synapse

cerebrum

the outermost germ layer of a human embryo, which eventually gives rise to the nervous system, skin, and sensory organs

the site of neuron communication between two cells, which permits one neuron to send an impulse to the next neuron

the dominant portion of the forebrain that is responsible for complex thinking, coordination of movement, and memory storage

myosin

isometric contraction

bursae

the protein found in the thick filament of sarcomeres that causes muscle contraction

the type of muscle contraction that causes increased muscle tension but does not result in skeletal movement

the fluid-filled sacs located in joint cavities that serve to cushion bone and reduce friction between a joint's moving parts

periosteum

thyroid

neurosecretion

the thin outer surface of a bone that contains an outer fibrous layer and an inner osteogenic layer

the endocrine gland located in the larynx that regulates metabolism in the human body

the secretion produced by nerve cells in order to stimulate the production and release of hormones in response to neural signals

memory cell

acquired immunity

follicle

the type of *B*-cell that stores information for producing antibodies in an immune response

the type of immunity that the body gains over time and exposure to specific antigens found on the surface of pathogens

the sac-like structure found within the dermis that produces hair

ceruminous gland

spleen

lymph

the gland that produces wax in order to protect the ear canal from foreign pathogens and other invaders

the abdominal organ that filters blood and produces white blood cells to help fight infection

the clear fluid that carries white blood cells throughout the body

What is the difference between the pulmonary circuit and the systemic circuit in the human cardiovascular system?

What is the purpose of the human cardiovascular system?

How does blood move through a four-chambered heart in a closed circulatory system?

In humans, the pulmonary circuit moves from the lungs to the heart in the pulmonary circuit, where it becomes oxygenated. Oxygenated blood then moves from the heart to the rest of the body's tissues via the systemic circuit.

The cardiovascular system transports oxygen and nutrients to body tissues, and carries carbon dioxide and wastes from body tissues for eventual excretion.

Blood flows from the right atrium to the right ventricle as the heart beats, opening the tricuspid valve and allowing blood to flow through. The right ventricle pumps the blood into the lungs for oxygenation. Oxygenated blood then returns to the heart and enters the left atrium before entering the left ventricle to be pumped to the rest of the body.

What is the primary organ of the cardiovascular system, and what is its function?

What are the stages of the cardiac cycle?

What occurs during the process of respiration?

The heart, which is an organ composed of cardiac muscle, is the primary organ of the cardiovascular system. Its primary function is to pump blood throughout the body through a series of contractions and relaxations.

The cardiac cycle consists of systole, which is the contraction of the heart that moves blood out of the ventricles, and diastole, which is the relaxation of the heart that allows blood to refill the atria.

During respiration, the body intakes oxygen from the atmosphere in order to break down glucose and provide energy to the body. Carbon dioxide, a byproduct of this process, is expelled after the exchange of gases. In humans, this process takes place in the lungs.

What is the function of the diaphragm in human respiration?

What role do alveoli play in gas exchange in the lungs?

How are the cardiovascular system and respiratory system interconnected?

The diaphragm is a muscle that separates the lungs from the abdomen in the thoracic cavity. During inhalation—the first phase of respiration—the diaphragm contracts to open up the thoracic cavity and allows air to flow into the lungs. During exhalation, the diaphragm relaxes and allows air to flow out of the lungs into the atmosphere.

Alveoli, which are air sacs found at the end of the smallest bronchioles branching throughout the lungs, contain a membrane that is only one cell thick. This thin layer allows oxygen to be diffused into the blood and carbon dioxide to be diffused out of the blood.

The respiratory system delivers oxygen to the lungs, where it diffuses in the blood and is sent to the heart to be transported to the rest of the body. The lungs also remove carbon dioxide from the blood that is pumped to the heart and expel the gas from the body through exhalation.

What is the difference between chemical and mechanical digestion?

What is the role of accessory organs in digestion?

Why is the human digestive system considered a complete digestive system?

Mechanical digestion is the physical breaking down of food. For example, the teeth and jaws physically chew food into smaller pieces for further digestion. Chemical digestion occurs when food is altered into different substances more suitable for absorption. Chemical digestion occurs when digestive organs secrete acids and enzymes that break down macronutrients.

Accessory organs—such as the liver and gall bladder—produce secretions (e.g., bile and chemicals) that complete chemical digestion. These organs are considered accessory because food does not directly pass through them as it is being digested.

In a complete digestive system, food is ingested at the mouth, travels through the digestive tract, and excreted at the anus. Vertebrates and complex invertebrates, such as arthropods, contain a complete digestive system. In contrast, an incomplete digestive system contains a digestive cavity with one opening for both ingestion and excretion. This is found among less complex invertebrates, such as cnidarians.

What is the function of villi in the small intestine?

What is the function of the large intestine?

What occurs in the process of defecation?

The villi, which are thin structures found within the small intestine, contain multiple folds. These folds increase the surface area of the organ and allow for more efficient nutrient absorption as food passes through the intestine.

The large intestine absorbs water and nutrients remaining in the digestive tract following their exit from the small intestine, then compacts and stores the remaining solid waste and moves it through the rectum for eventual excretion.

Defecation is the final process of digestion, in which solid waste is expelled from the digestive tract. Peristaltic waves in the large intestine move waste to the rectum, where it is held until pressure pushes it out of the anal canal.

What role do kidneys play in osmoregulation?

What path does urine take as it exits the body through the urinary tract?

How does the male urethra function in both the urinary system and reproductive system?

Osmoregulation is the process of balancing the amount of water and salt in bodily fluids, such as blood. Kidneys are responsible for filtering waste from the blood; as blood filters through the kidneys, it maintains osmoregulation through the kidney's excretion of excess water or conservation of water when there is too little of it in the bloodstream.

Urine is created in the kidneys and travels down long, thin ureters to the bladder for storage. The stored urine is eventually released out of the body through the urethra.

In males, urine is excreted out of the body through the urethra as the urethra relaxes and urine stored in the bladder relaxes, which pushes urine out of the body. The male urethra is also the site of ejaculation, during which a mixture of sperm and fluids, called semen, is expelled during intercourse.

Why do gametes—specialized reproductive cells created by sexually reproducing animals—have half of a full set of chromosomes?

Where does oogenesis occur in the female reproductive system?

What are the two main parts of a male sperm cell, and what are their functions?

Sexual reproduction requires one set of genes from each of the two parents. Female gametes (eggs) and male gametes (sperm) are haploid, meaning they have half of a full set of chromosomes. When fertilization occurs, genes from both parents recombine to create genetically distinct offspring.

Oogenesis—the production of eggs—initially begins in the outer layers of the ovaries and matures in the follicles during adolescence.

The head of the sperm cell contains the nucleus as well as an acrosome, which holds enzymes to penetrate a female egg. The tail of the sperm is called a flagellum, which undulates back and forth to create movement.

What are the stages of human embryogenesis?

What are the parts and functions of a neuron, or nerve cell?

What distinguishes the central nervous system and peripheral nervous system from each another?

After fertilization, the newly formed cell, or zygote, divides rapidly to become a blastula during the blastulation process. The cells within the blastula rearrange and start to specialize into distinct germ layers during gastrulation. These germ layers will eventually become specialized tissue and organ systems.

Neurons are comprised of a cell body, dendrite, and an axon. Cell bodies contain the nucleus; dendrites receive impulses from other neurons; and axons transmit impulses across a synapse to another neuron.

The brain and the spinal cord make up the central nervous system and are primarily responsible for processing and integrating external information and coordinating activity throughout the body. The peripheral nervous system is a system of nerves that is responsible for transmitting information back and forth from the various parts of the body to the central nervous system.

What are the components of a reflex arc?

What are the components and function of the brain stem?

Which type of muscle is primarily responsible for all voluntary action in the human body?

A reflex arc is composed of a sensory neuron and a motor neuron. During an involuntary reflex action, the sensory neuron synapses in the spinal cord and activates the motor neuron without first traveling to the brain. This results in a nearly instantaneous, involuntary motion in response to a stimulus.

The brainstem is comprised of the medulla oblongata, pons, and midbrain. Collectively, these components connect and transmit sensory and motor messages between the brain and the spinal cord. The brainstem also controls many autonomic activities, such as heart rate and breathing.

Skeletal muscles are the only muscles in the body that are controlled consciously. The other types of muscle—cardiac and smooth—are involuntary muscles that control autonomic movement within organ systems.

How do skeletal muscles move?

What is the difference between fast-twitch fibers and slow-twitch fibers?

How does the cardiac muscle tissue control contractions of the heart?

Skeletal muscles are typically attached to two bones across a joint. They are anchored by tendons, or dense bands of connective tissue. When muscles contract, they pull on the tendons and move them closer to one another to create skeletal movement.

Type I, or slow-twitch, muscle fibers are fibers that derive energy from aerobic respiration in order to make slow, repeated movements over long periods of time. Type II, or fast-twitch, muscle fibers derive energy from anaerobic respiration in order to produce fast movements for a short period of time.

Cardiac muscle tissue contains specialized cells, called pacemaker cells, which produce electrical charges that stimulate a heartbeat by continuously polarizing and depolarizing it, causing a chain reaction that moves electricity from cell to cell.

What role does bone marrow play in blood production?

How are tendons, joints, and ligaments connected and interrelated?

What are the major bones of the axial skeleton?

Red bone marrow found in some of the human body's bones is the site of blood cell production, or hematopoiesis. The hematopoietic stem cells produced in this marrow create white blood cells, red blood cells, and platelets.

Ligaments, which are bands of flexible tissue, connect two bones together to form a joint. Tendons are flexible tissue bands that attach muscle to two bones that form a joint. Muscles contract to move the bones that are part of the joint, and ligaments protect the bones and joint through this movement.

The axial skeleton protects the vital organs at the center of the human body. It is composed of the skull, which includes cranial, facial, and associated bones; the thorax, which includes the sternum and ribs; and the bones associated with the vertebral column.

What minerals are stored in the human skeletal system?

What are the functions of the endocrine system?

What differentiates the posterior pituitary gland from the anterior pituitary gland?

The bones found in the human skeletal system are a major storage center for calcium and phosphorus and account for the vast majority of these minerals in the body. Magnesium and fluoride are also found in the bones.

The endocrine system secretes hormones into the body to regulate and control major bodily functions, such as metabolism and growth, and respond to stimuli in order to maintain homeostasis.

The posterior pituitary gland, located in the back, receives then transmits hormones produced by the hypothalamus. The anterior pituitary gland, located in the front, produces and releases its own hormones.

How do hormones transmit chemical messages?

What role does the hypothalamus play in homeostasis?

Why are physical barriers, such as cilia, mucus, and saliva, considered nonspecific defenses of the immune system?

After hormones are secreted and released into the bloodstream, they bind to receptor molecules extending from target cells. Then, the hormones alter the cell's production of proteins, enzymes, and other structures to stimulate the intended change.

The hypothalamus is the portion of the brain that serves as the main link to the endocrine system. It maintains homeostasis by regulating the release and inhibition of hormones throughout the body.

Nonspecific defenses are the immune system's way of targeting any pathogen that could potentially pose a threat to the body's well-being. Cilia, mucus, and saliva do not target a specific pathogen; instead, they form a barrier between any foreign object and the body.

What is the difference between the adaptive immune system and the innate immune system?

How do helper *T*-cells coordinate immune responses to pathogens?

What occurs during the process of phagocytosis?

The innate immune system uses nonspecific defenses to prevent foreign matter from entering the body and attacking any foreign cell that does enter the body. In the adaptive immune system, there is an immune response that recognizes and responds to pathogens that enter the body based on their specific properties.

Helper *T*-cells are a type of lymphocyte that seek out and bind to specific antigens. Once they are bound to the antigen, they stimulate one of two responses: a cell-mediated response or an antibody-mediated response.

During phagocytosis, the phagocyte cell adheres to the foreign or invading particle. The particle is then engulfed within a vacuole, or phagosome, and is digested by enzymes carried by lysosomes within the cell.

What are the different layers of the dermis, and what are their functions?

Why do sudoriferous glands in the dermis produce sweat?

What role does the integumentary system play in vitamin *D* synthesis?

There are two distinct dermis layers of the skin: the papillary and the reticular. The papillary layer delivers blood and nutrients to the epidermis and contains nerve cells to receive messages from outer stimuli. The reticular layer is made of collagen and elastin fibers, which give skin its strength and elasticity.

Sudoriferous glands produce sweat, which travels to the skin and evaporates in order to lower the body's temperature. Waste materials are also excreted through sweat as a way to reduce their presence inside the body.

Cells within the epidermis produce vitamin *D* in response to exposure to ultraviolet light, which is received from sunlight. The skin also functions to store vitamin *D*.

What is the function of keratinization?

What occurs during a negative feedback loop in the human body?

How do the endocrine, cardiovascular, and lymphatic systems work together to fight disease?

Keratinization is the hardening process of the keratin protein in skin, hair, and nail cells. These hardened cells help protect the skin, hair, and nails from damage and foreign particles.

Feedback loops play a major role in maintaining homeostasis. During a negative feedback loop, which is the most common loop, the body responds to an external change by eliciting a response to counteract the change and return the body to homeostasis.

White blood cells are produced in response to chemical stimuli sent to the target organ by the endocrine system. They are transported to the site of infection or site of a foreign invader via the vessels of the cardiovascular and lymphatic systems.

What types of substances are transported by the lymphatic system?

What is the function of lymphatic capillaries?

The lymphatic system transports many different types of interstitial fluid, including water, waste materials, and hormones. It delivers white blood cells to sites of damage or infection and fatty acids from the digestive system to the cardiovascular system.

Lymphatic capillaries are small vessels of the lymphatic system that absorb excess fluids from tissues throughout the body in order to maintain a balanced fluid level in the body.

Made in the USA
Middletown, DE
10 July 2017